JAMAICA

AF775324

PEP

Mathematics

Performance Task

GRADE 6

Practice Papers

hachette LEARNING

The Publishers would like to thank the following for permission to reproduce copyright material.

Photo credits

t = top, b = below, l = left, r = right, c = centre, b/g = background

p.14: © TWINS DESIGN STUDIO/stock.adobe.com; p.16: © TWINS DESIGN STUDIO/stock.adobe.com; p.22: © xenial/stock.adobe.com; p.31: © Comauthor/stock.adobe.com; p.33: © Yael Weiss/stock.adobe.com; p.60: © Milan/stock.adobe.com; p.66: (l) © Alena/stock.adobe.com, (r) © Alena/stock.adobe.com; p.67: © Alena/stock.adobe.com; p.68: (l) © Production Perig/stock.adobe.com, (c) © showcake/stock.adobe.com, (r) © Savvapanf Photo ©/stock.adobe.com; p.71: © zelimirzarkovic/stock.adobe.com; p.77: © Karol Kozlowski/alamy.com.

Acknowledgements

Although every effort has been made to ensure that website addresses are correct at time of going to press, Hachette Learning cannot be held responsible for the content of any website mentioned in this book. It is sometimes possible to find a relocated web page by typing in the address of the home page for a website in the URL window of your browser.

Hachette UK's policy is to use papers that are natural, renewable and recyclable products and made from wood grown in well-managed forests and other controlled sources. The logging and manufacturing processes are expected to conform to the environmental regulations of the country of origin.

To order, please visit www.HachetteLearning.com or contact Customer Service at education@hachette.co.uk / +44 (0)1235 827827.

ISBN: 9781036005214

First published in 2026 by Hachette Learning (a trading division of Hodder & Stoughton Limited),
An Hachette UK Company
Carmelite House
50 Victoria Embankment
London EC4Y 0DZ

www.HachetteLearning.com

The authorised representative in the EEA is Hachette Ireland, 8 Castlecourt Centre, Dublin 15, D15 XTP3, Ireland (email: info@hbgi.ie)

Impression number 10 9 8 7 6 5 4 3 2 1

Year 2030 2029 2028 2027 2026

Cover photo © GoodStudio- stock.adobe.com

Illustrations by Hyphen S.A./Konstantina Kyratzidou

Typeset by Hyphen S.A.

Printed and bound in Great Britain by Bell & Bain Ltd, Glasgow

A catalogue record for this title is available from the British Library.

FSC
www.fsc.org
MIX
Paper | Supporting responsible forestry
FSC™ C104740

Contents

How to use this book

This term you will write your final maths performance task. The information in this section and the six new practice tasks in this book are all designed to help you prepare for this task and to make sure that you know what to expect and how to answer questions so that you can do your very best.

What to expect …

You did performance tasks in Grade 4 and Grade 5. The Grade 6 performance task will be similar. That means:

- You will have to answer questions about a topic from everyday life.
- There will be four or five parts to the task.
- Information that you need to answer the questions will often be given in words, tables and diagrams, so you need to read these carefully.
- Some questions will be open-ended. That means there is more than one right answer, and you are asked to make decisions and show what you think the answer is.
- You will get credit for showing your ideas and your working out even if you do not have the correct answers.

Performance tasks are different to multiple choice curriculum tests or ability tests. The aim is for you to apply what you know and use your reading, maths and thinking skills to answer the questions. To get a good result, you must show clear evidence of your learning. You cannot just guess the answer or circle the correct choice.

Your performance task will be done in class at a specific time. You must finish it in class and you cannot take it home to complete it.

How can you prepare for a performance task?

You cannot study for a performance task in the same way that you might study for a maths test. This is because you do not know beforehand what topics will be covered in the task or what you are going to be asked to do. What you do know is that the task will focus on things you have already learned. You will not be expected to do any unfamiliar maths.

Even though you do not need to study for a performance task, there are many things that you can do to prepare yourself and make sure you do well.

Set realistic goals

Think about what you would like to achieve. Your work will be assessed and rated using a four-level scale.

1 Beginning	2 Developing	3 Proficient	4 Highly proficient
Answers show no evidence or very little evidence that the student understands the concepts.	Answers show some evidence that the student partly understands the concepts.	Answers show that the student is competent and understands most of the concepts.	Answers show that the student understands all the concepts very well and may have found the task quite easy.

Ideally, you should aim to be rated 3 or 4 on the task.

Revise by working through model tasks

There are six model tasks in this book. You can use these to prepare for your in-school performance task.

- Work through the task on your own or with a partner. Remember that working in pairs or groups can help you learn.
- Check your own answers.
- Reflect on your work. Identify what you did well and which areas you need to improve. Decide what you will do to make the improvements you need.

The flow diagram shows the process of revising using the model tasks.

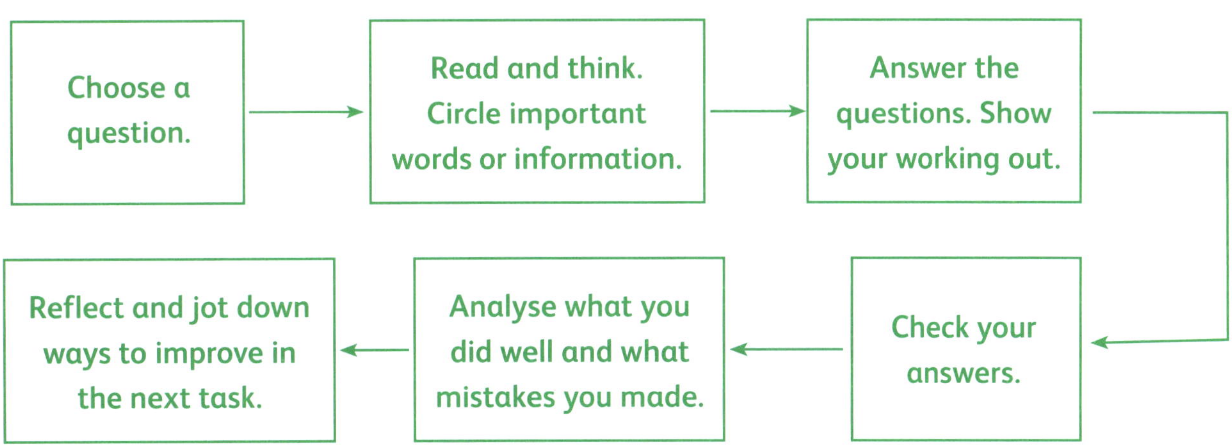

When you are preparing, you do not need to work through the entire task at one time, you can do one or two questions. It is important to try to do at least one whole task to help you develop good techniques.

Developing good techniques

Practice tasks can help you develop some good techniques for managing your in-class tasks. These include:

- managing your time
- breaking down the task into manageable pieces
- reading and making sense of questions and what you need to do
- choosing suitable strategies for different types of questions
- developing clear ways of showing your working out and thinking
- writing your answer clearly so that it makes sense.

Practicing and learning how to manage tasks can also help you feel less anxious and pressurised because you are prepared and you know how to work through any type of task.

Well-being tips

Stress and pressure can affect your performance and, if you are tired, you are more likely to make careless mistakes. It is important that you take good care of yourself and that you eat well and exercise while you are preparing for tasks and tests.

The night before a task or test, try to get a good rest.

Get up on the morning of the test and get ready for the day. Have a healthy breakfast and make sure you are organised and that you have the things you will need (pencil, ruler, eraser and calculator).

Try to get to school on time so you do not need to rush.

Stay positive and remember that you are well-prepared and that you know what is expected.

Read these thoughts. Do you ever think like this?

I'm really scared that I will not do well on this task.

I feel like everyone else is going to do well and that I am not.

I hate doing tests, I always freeze up.

I just hope I can work fast enough to answer all the questions.

What am I going to do if I do not understand the question?

Everyone expects me to do brilliantly. What if I do not?

I rushed this morning and now I feel all flustered.

When you feel anxious or stressed out, you can try a technique called box-breathing. Fire-fighters and other emergency workers use this technique to help them remain calm in difficult situations.

Step 1: Breathe in, counting to four slowly. Feel the air enter your lungs.

Step 2: Hold your breath for four seconds. Try to avoid inhaling or exhaling for four seconds.

Step 3: Slowly exhale through your mouth for four seconds.

Step 4: Wait four seconds before inhaling again.

Repeat the steps till you feel calm.

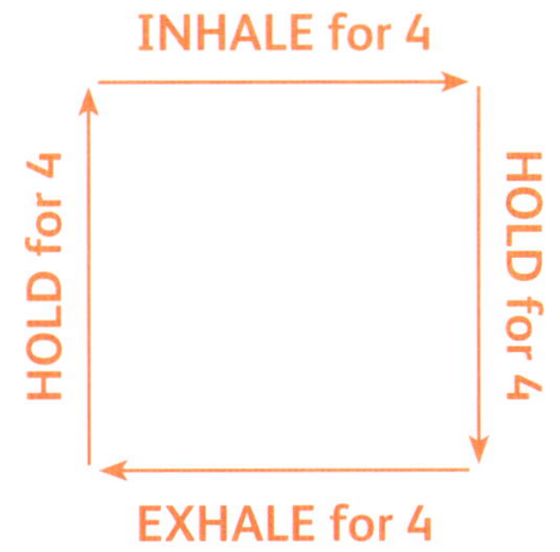

Tips for answering questions

Here are some general guidelines for answering questions in your performance task and other tests.

Before the task …

Listen carefully to all instructions from your teacher.

Take note of how long you have and jot down what time you need to finish.

Double check that you have the equipment you need.

Ask questions if anything is unclear.

Getting started …

Skim read all the questions before you start.

Decide on the order in which you will answer the questions.

Work carefully and systematically.

Write your answers in the spaces provided.

Making sense of the questions ...

Take time to understand the question.

Look for maths terms and remember what they mean.

Find details in the problem and circle or highlight them.

Plan how you will answer it (strategy).

Work out the answer, showing all steps.

Check your answer by reading the question again.

Why is it important to show your working out?

Read these students' answers for each part of the question.

Both students have the wrong answers, but Nadia gets a better grade than Annika because the teacher can see some evidence that Nadia understood the question and that she used the correct methods, even though she got the answers wrong.

A taxi travels 380 km in five hours.

a What is its speed in km/hr?

Nadia	Annika
Speed = distance ÷ time 380 ÷ 5 = 75 km/hr	75

b How long would it take the taxi to travel 532 km at this speed?

Time = distance ÷ speed 532 km ÷ 75 = 7.09 It will take 7 hours and 9 minutes	About 7 hours

Different questions need different types of answers

When you read through the task, you will have to decide for yourself how to answer each question and each part of a question.

This section gives you some guidelines on answering different types of questions.

Remember to start by reading the task carefully.

You will see that some questions are quite simple, and you can answer by writing and showing a calculation. Others are more complex, and you need to show different steps in the process.

Performance task: Planning a class party

Class 6 is going to have a party. The bar graph shows the amount they have budgeted for different party items.

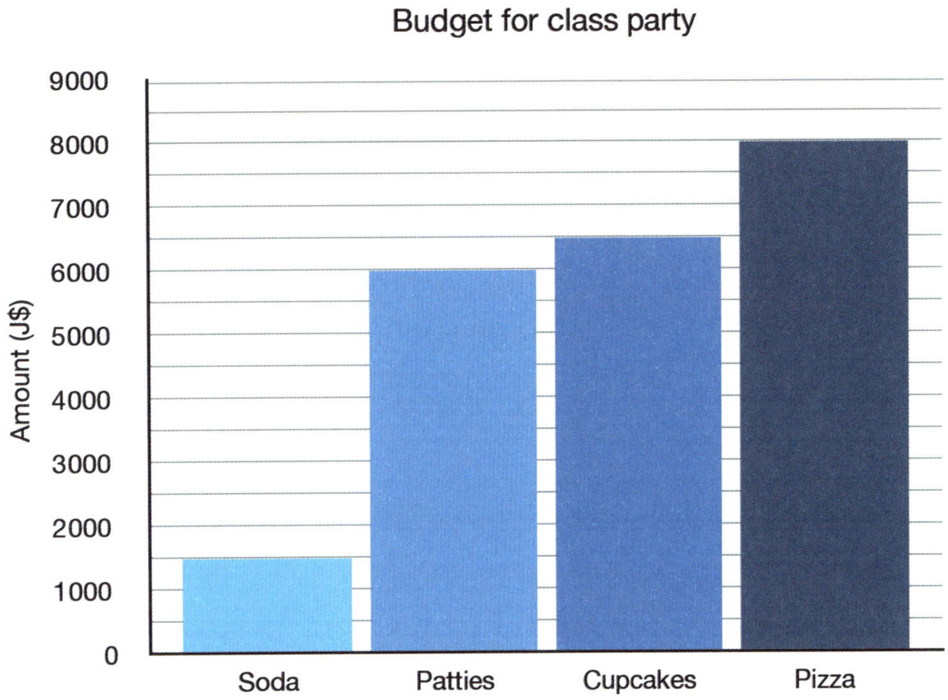

This tells you that you will need to use the bar graph to find the answers.

Use information from the bar graph to complete the task.

1 How much money did the class budget for the party? Show how you worked out the answer.

These amounts come from the graph. You need to add them all up to find the total amount budgeted.

Always write the answer.

$1 500 + $6 000 + $6 500 + $8 000 = $22 000

The class budgeted $22 000 for the party.

2 There are 24 students in the class. The teacher wants to buy one cupcake for each student.

Use the information you are given.

Show your calculation.

Explain in words, how you thought about the problem, and how to solve it.

a How much can the teacher spend on each cupcake? Explain your reasoning.

Budget for cupcakes: \$6 500

Number of students: 24

\$6 500 ÷ 24 = \$270.83

The graph shows that they have \$6 500 to spend on cupcakes. I divided that among the students and got \$270.83. That means they can spend any amount up to \$270.83 per cup cake.

Show each step.

Show that you have added the two amounts.

Remember to write an answer.

b The teacher ends up buying 24 cupcakes that cost \$265 each and 24 patties that cost \$240 each. What did he spend on these two items?

24 × \$265 = \$6 360

24 × \$240 = \$5 760

\$12 120

He spent \$12 120 on these two items.

Use the information from the graph. Work out how many they can buy.

Show how you work out the number of cups.

Write the answer.

3 It costs \$300 for a 2-litre bottle of soda. If the class spend their entire soda budget on the 2-litre bottle, how many 250 ml cups can they pour?

Budget is \$1 500.

5 × \$300 = \$1 500, so they can buy 5 bottles.

5 × 2 litre-bottles = 10 litres of soda

You can get 4 × 250 ml cups from 1 litre.

10 × 4 = 40 cups

They can pour 40 cups.

4 The teacher thinks that three medium pizzas will be enough for the party. The class find out the prices of pizzas from four different pizza places.

Pizza prices (medium)

Pizza Mania	$2 550	**Mama's Pizza**	$2 620
Pizza Palace	$2 790	**Little Dom's**	$2 690

> You are being asked for a recommendation and there is more than one correct answer.
> Show clearly how you think about the problem.

a Which pizza would you recommend they buy? Why?

They have $8 000 for pizza.

$8 000 ÷ 3 = $2 666.67

The most they can spend is $2 666.67 per pizza, so they should buy from Pizza Mania or Mama's Pizza because those are both within their budget. They should choose the one that is closest to them.

> This is also an open-ended question, and you have to think creatively and use what you have already worked out to solve it.

b Some students want to order from Pizza Palace because the pizzas there are very cheesy. How could they do this without going over the class budget?

Three pizzas from Pizza Palace would be

3 × $2 790 = $8 370

This is $370 more than they have in the pizza budget. They spent less than they budgeted on cupcakes and patties, so they have $380 left over from those items. That is enough to cover the cost of the pizza.

The class could also buy fewer bottles of soda. They only need 24 cups of soda, which is 6 litres, so if they only buy 3 × 2-litre bottles of soda, they would spend $900 and have $600 over to add to the pizza budget.

> Give clear reasons, with numbers, where possible.
> Try to think of more than one way to do this.

c Show how you could share the three pizzas so that each student gets an equal share. <u>What is each student's share?</u>

$\frac{\text{3 pizzas}}{\text{24 students}}$

3/24 = 1/8

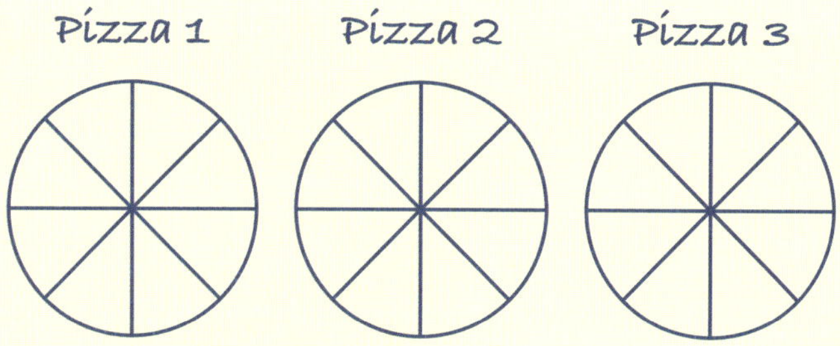

24 ÷ 3 = 8, we need 8 slices from each pizza.
Each student will get one slice which is 1/8 of a pizza

This is a good place to show your answer as a drawing.

Remember to write the share as a fraction of a pizza.

There are more worked examples showing both good and poor answers at the end of **Practice paper 1**.

Practice paper 1

Always check that you have everything you need before you start.

This task has 3 parts:

- Part 1 has questions 1 and 2.
- Part 2 has questions 3 and 4.
- Part 3 has question 5.

You will need a calculator, a pencil and an eraser.

Answer all the questions in each part in the spaces provided.

Remember to always show your working out.

→ Before you start …

The instruction words have been underlined in this sample task. Think carefully about what each instruction means before you answer the question.

These are some of the instruction words and phrases you will find in this task.

Calculate …	How can you …	Show …
Choose …	How does … compare …	
Compare …	How many …	Show how you …
Decide …	How much more …	What fraction …
Explain why …	Justify …	What is …
Fill in …	Round …	Work out …

Colour the words to match what you need to do for each.

● Recalling information	■ Skills and concepts	▲ Strategic thinking	★ Extended thinking
Remember a fact or way of calculating what you have learned and write it down. There is **one** correct answer.	Use your skills and what you know to make decisions about what to do and how to do it. There is **one** correct answer.	Explain your thinking and give reasons for your answers. There are different answers and different ways to find the answers.	Plan, think and put ideas together to solve a problem. There are many possible answers and ways of solving the problem.

Part 1

The smoothie store

Super Smoothies has this information on display in the store.

Read all the information on the poster. You will need to find information about sizes and prices so you can answer the questions.

SUPER SMOOTHIES

FAB FLAVOURS

Very Berry

Mango Madness

Green Goodness

Papaya Power

Pine + Ginger

Join us for $50 off Fridays

max 4 per customer

Add a scoop of sorbet $75

350 ml $700

500 ml $850

OPEN: Mon–Fri 9 a.m.–6 p.m. Sat + Sun 9 a.m.–9 p.m.

Question 1

Use the information on the poster. Write a number sentence to show how you work out each answer.

a What is the difference in the size of the two smoothies?

What operation do you need to find the difference?

This is another way of asking you to find a difference.

b How much more expensive is the large smoothie compared with the medium?

How does a shop work out what you need to pay?

c If you buy two medium smoothies and add a scoop of sorbet to one of them, how much will you pay?

Why does this question refer to a Friday? Look back at the poster to check.

d If you buy five large smoothies on a Friday, how much will you pay?

Where will you find the numbers you need to do the calculation?

e The table shows the number of smoothies sold each day last week at Super Smoothies.

Day	Number sold
Mon	42
Tues	38
Wed	45
Thu	41
Fri	68
Sat	72
Sun	65

What is the mean (average) number of smoothies sold per day? ______

Choose any smoothie price (e.g. medium at $700).

Calculate the cost of 4 smoothies with the holiday deal.

Calculate the cost of 4 smoothies on $50 off Friday.

Which total is lower?

f During the school holidays, Super Smoothies offers a special deal. If you buy four smoothies, you get the cheapest one at $\frac{1}{2}$ price. Jerome says you can pay less for four smoothies if you buy them on $50 off Friday.

Compare the two offers. Decide whether Jerome is correct or not. Explain your answer and show the calculations you used to make your decision.

Question 2

Super smoothies uses these ingredients to make a medium Mango Madness smoothie.

Mango Madness

100 g mango

50 g pineapple

100 g banana

250 ml fresh orange juice

The word "what" often means you have to do a simple calculation to find the answer.

a What fraction of the ingredients is orange juice?

Do not just write the amounts. Remember to show how you work them out.

b What is the ratio of mango to pineapple to banana in the smoothie?

c To make a large Mango Madness, they use 50% more of each ingredient. Work out how much of each ingredient they will need to make the large smoothie.

_______________ mango

_______________ pineapple

_______________ banana

_______________ orange juice

Write the medium ratio (part b).

Write the large ratio (part c).

Can they simplify to the same?

What does that show?

d Is the ratio of mango to pineapple to banana the same for a large smoothie? Justify your answer.

Part 2

Comparing prices

Question 3

Marcia finds this box of ready-made smoothies for sale at the local supermarket for $1 999.

Courtney finds this box of ready-made smoothies for sale at the supermarket for $2,000.

Remember:
1 kg = 1,000 g.

a What is the mass of one bag of smoothies in grams?

To find the price for 1 kg, divide the total price by the total kilograms.

b What is the price per kilogram?

c Another brand costs $1,600 per kilogram. Which brand offers a better deal?

How do you normally answer the question, "What does it mean?"

d What does "5 pre-portioned 250 g bags" mean?

There are two instructions in this question.

e What is the price per bag if you buy this box? Round your answer to the nearest dollar.

Use your own answer, even if it is not correct, and the information on the poster, to compare the prices.

f How does the price of these bags of smoothies compare with the prices at Super Smoothies?

Question 4

A customer writes a review of Super Smoothies and complains that the smoothies are too expensive. The owner of Super Smoothies writes this response to the customer review.

Super Smoothies

Hi Jamaal,

Thanks for your review. I just want to point out that we use only fresh ingredients in our smoothies and that they are actually good value for money.

Concentrated orange juice is 40% cheaper than the 100% fresh juice that we make.

Concentrated lime juice is 30% cheaper than using fresh limes as we do.

I hope that clears things up for you.

We hope to see you in store soon.

a The customer decides to check this. He finds out that 1 litre of orange concentrate is $1 272 at the store. He finds out that it takes eight oranges to make 1 litre of juice and that oranges cost about $245 each.

Fill in the missing numbers in Jamaal's calculations.

Price concentrate = $1 272

Price fresh = ________ × $245 = $1 960

Concentrate as percentage of fresh juice

$1 272/$1 960 × ________ = 64.89%

That means fresh juice is only 35.11% cheaper, so the owner is wrong.

Show how you work out your answers.

You have to show **two** things here. Think carefully about how to do this before you answer.

b Show how the owner of Super Smoothies could be both correct and incorrect using those figures.

What fraction of a litre is 750 ml? Could this help you find the answer?

c The price of 750 ml of store-bought concentrated lime juice is $265. Work out the cost per litre.

How do you compare amounts in maths? Could you use symbols?

d Compare this to what it would cost Super Smoothies if they make a litre of juice using fresh limes.

Part 3

Management decisions

Question 5

Super Smoothies get a lot of requests for small smoothies and extra-large smoothies, so they decide to add these sizes.

You can choose any sizes as long as you explain why you have chosen them.

a Choose a suitable size for each new smoothie in millilitres. Explain why you chose each size.

____

____

____

____

Remember the amount of each ingredient was already given for the medium smoothie in Part 1, question 2. Go back and find them to use them for this question.

b Use the sizes you have chosen and work out how much of each ingredient you will need to make a Mango Madness smoothie in each size.

Small Mango Madness	X-Large Mango Madness
______ Mango	______ Mango
______ Pineapple	______ Pineapple
______ Banana	______ Banana
______ Orange juice	______ Orange juice

Show your working out.

"Decide" means that you make the choice. Any answer you give will be correct as long as you explain how you made up your mind.

c Decide on a selling price for each new size. Explain how you worked this out.

d (Optional challenge question):
Super Smoothies surveys 120 customers about new sizes:

- 45 want small
- 50 want medium
- 25 want extra-large

They can only make 200 smoothies a day. Based on this information, how many should they make of each size daily?

Practice paper 1 Self-evaluation

How do you know how well you did?

Now that you have worked through the task, you are going to look at **some** answers that students gave to see how they were rated. Then you can rate your own answers using this scale.

1 Beginning	2 Developing	3 Proficient	4 Highly proficient
Answers show no evidence or very little evidence that the student understands the concepts.	Answers show some evidence that the student partly understands the concepts.	Answers show that the student is competent and understands most of the concepts.	Answers show that the student understands all the concepts very well and may have found the task quite easy.

We will look at answers to the four different levels of questions.

Recalling information	**Skills and concepts**	**Strategic thinking**	**Extended thinking**

Recalling information

Question 1(a)

a What is the difference in the size of the two smoothies?

One is bigger	500 ml − 350 ml = 150 ml The large smoothie contains 150 ml more.
1 Beginning *The student does not show that they understand the term "difference" and hasn't done any maths to answer the question.*	4 Highly proficient *The student has done a calculation to find the difference and has also written a statement as an answer.*

Question 2(b)

b What is the ratio of mango to pineapple to banana in the smoothie?

100 : 50 : 100	100 : 50 : 100 = 10 : 5 : 10
② Developing *The student understands the concept of ratio and has compared the amounts correctly but has not simplified the ratios nor indicated what they are comparing.*	③ Proficient *The student understands the concept of ratio and has simplified the ratios but not indicated what they are comparing, for example mango : pineapple : banana = 10 : 5 : 10.*

Skills and concepts

Question 3(b)

b What is the price per kilogram?

You know it costs **$1999 for 1.25 kg**, so you can divide to find the cost of 1 kg.	I would use a bar model like this: 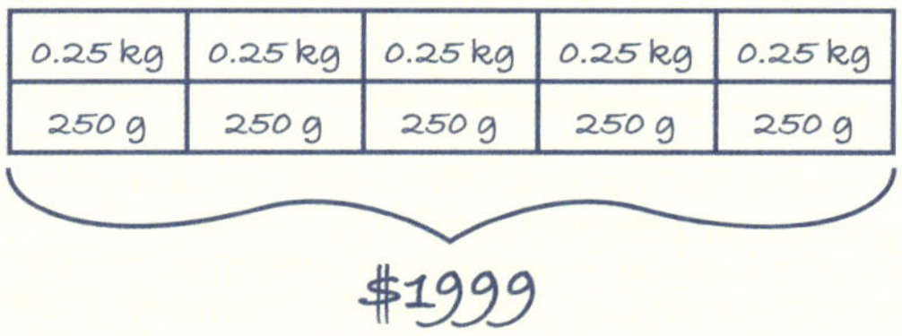 $1999 ÷ 1.25 = 1599.20 per kg
③ Proficient *The student has answered the question partially and shows they understand they need to divide to find the price per kilogram. However, they have not explained **what they are dividing by** or **shown their working**.*	④ Highly proficient *The student shows a good understanding of what is being asked and has used a mathematical model to show how they would solve the problem. The model indicates that they **understand the relationship between cost and weight** and can **apply division to find a unit rate**.*

Question 2(d)

d Is the ratio of mango to pineapple to banana the same for a large smoothie? Justify your answer.

It is 150 : 75 : 150.	The ratio of mango : pineapple : banana in a medium smoothie is 10 : 5 : 10. The ratio of mango : pineapple : banana in a large smoothie is 150 : 75 : 150. If you divide each by 15, you get 10 : 5 : 10. So, the ratio is the same.
② Developing *The student has worked out the ratio for the large smoothies but not answered the question and has not justified their answer.*	④ Highly proficient *The student has compared the ratios and then shown by simplifying (efficiently) that they are equivalent.*

Strategic thinking

Question 1(f)

f During the school holidays, Super Smoothies offers a special deal. If you buy four smoothies, you get the cheapest one at $\frac{1}{2}$ price. Jerome says you can pay less for four smoothies if you buy them on \$50 off Friday. Compare the two offers. Decide whether Jerome is correct or not. Show what maths you do to make your decision.

If you buy four small smoothies, it costs \$2800 (4 × \$700). On a Friday, this costs 4 × \$650 = \$2600. If you get one half price, then you pay 3(× \$700) + 350 = \$2450. This is cheaper than a Friday, so Jerome is wrong.	Half of a small is \$700 ÷ 2 = \$350. Half of a large is \$800 ÷ 2 = \$400. You will either get \$350 or \$400 off the price if you take the special offer. The most you can get off on a Friday is 4 × \$50 = \$200. This is less than the half price of any smoothies, so Jerome is wrong.
③ Proficient *The student has compared the offers using one set of prices and shown the maths that they used to make their decision. They have not fully shown that Jerome is wrong for all combinations of smoothies.*	④ Highly proficient *The student's answer shows that they can think in general terms and find a solution that works for any combination of smoothies. They have given a simple, concise answer to the problem.*

Question 4(d)

d Compare this to what it would cost Super Smoothies if they make a litre of juice using fresh limes.

This is cheaper.	Cost of concentrate juice is \$353.33/litre. The owner says this is 30% cheaper than fresh. So, fresh would cost: \$353.33 ÷ 70 × 100 = \$504.76 \$504.75 − \$353.33 = \$151.42 The concentrate juice is \$151.42 cheaper per litre.
① Beginning *The student has made a statement but shown no comparison to support it. This could just be a guess or an acceptance of the statement made by the owner.*	④ Highly proficient *The student has shown that they understand how to work with percentages and has shown both prices to compare them, and given a detailed answer.*

Extended thinking

Question 4(b)

b Show how the owner of Super Smoothies could be both correct and incorrect using those figures.

The owner is wrong because she said it costs 40% more and Jamaal showed that it was only 35.11% cheaper.	If you round to the nearest 10%, then 35.11% rounds up to 40%, so if the owner rounded like that, she would be correct to say 40%. If she rounded to the nearest per cent, then the difference is only 35% and she would be incorrect.
2 Developing *The student has shown that they understand the comparison, but they have accepted the numbers at face value and not answered the question to show "how"the owner could be correct as well.*	4 Highly proficient *The student shows that they understand how numbers (percentages) can be used in different ways and how rounding to different places can affect the values you use. The answer shows that the owner can be both correct and incorrect.*

Question 5(c)

c Decide on a selling price for each new size. Explain how you worked this out.

Small \$600. X-Large \$1 000.	My small size holds 25% less than the medium, so I would charge \$700 − 25% = \$525 for the small. My X-large holds 40% more than the large. That works out to \$1190, but that is an odd price and you'd have to keep giving \$10 change, so I would sell the X-large for \$1200.
2 Developing *The student has given prices but no indication of how they worked these out.*	4 Highly proficient *The student has made decisions on pricing based on the sizes they chose earlier and has explained how they thought about the proportions to create them. They have also indicated awareness of rounded pricing to avoid change, making it easier to work with.*

Reflection

Go back through your own answers to the questions in this task.

- Which ones did you do well?
- What made these ones easy for you?
- Which ones did you struggle with?
- What made them difficult for you?
- What two things can you do to make these types of questions easier to work through in the future?

Practice paper 2

This task has 4 parts:

- Part 1 has questions 1 and 2.
- Part 2 has questions 3, 4 and 5.
- Part 3 has question 6.
- Part 4 has question 7.

Read all the information in each part carefully. You will need to use the information to answer the questions.

You must answer all the questions in each part.

→ Before you start …

Read through all the questions.

Ask yourself, "What am I being asked to do here?"

Jot down your ideas next to each question.

Here are some key words that might help you.

Use a formula. Do a calculation.
Solve a two-step problem. Draw a graph.
Show working on a diagram. Use patterns.
Give an opinion. Make decisions. Give reasons.

Road trip!

Highlight all the numbers and measurements in the information so you can find it easily if you need it to answer questions.

Kwacy is nine years old. He is excited to go on a road trip with his dad, Steve, his mom, Rebecca, and his little sister, Kalanda, who is three years old.

The family will drive from Super Town where they live, all the way to NuTime City which is 81 km away. Kwacy is looking forward to all the fun places he and his family will visit on their road trip.

Part 1

BestCo gas station

Question 1

The car gas tank has a capacity of 54 litres. The tank is now one-sixth ($\frac{1}{6}$) full. The family needs a full tank to go on the trip.

Dad asks the pump attendant to fill his car tank with gasoline 90.

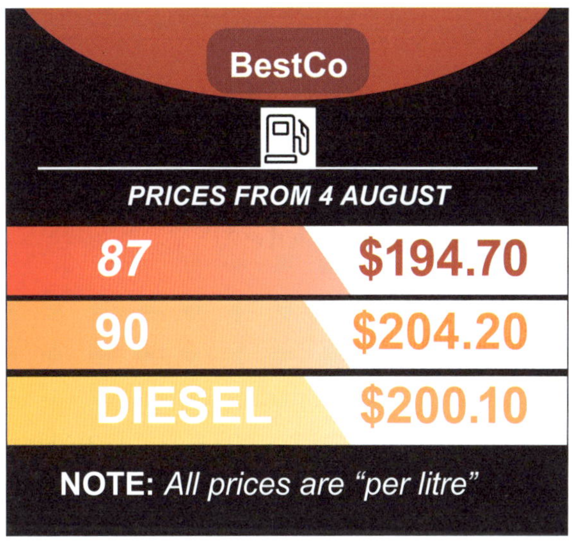

Think: What information do I need?

Tank capacity: ____ litres

Current amount: 1/6 fulll

Operation: ____ × ____

a How much gas (in litres) is in the tank before the attendant fills it up?

b What is the difference between the price of gasoline 90 and gasoline 87?

First find how much gas is needed to fill the tank. Then multiply by the price per litre.

c How much will it cost to fill the tank?

Question 2

Dad has $20 000 in his wallet:

- one $5 000 note
- five $2 000 notes
- five $1 000 notes.

Dad asks Kwacy to take money from his wallet to pay the pump attendant.

Start with the largest note value. Can you make the total using fewer notes this way?

a What is the fewest notes that Kwacy can take out of the wallet to pay for the gas?

b How much change should Kwacy receive?

Part 2

"USD" means "United States Dollars".

At the *Fun-Funn* Adventure Park

Read this information about the *Fun-Funn* Adventure Park.

Fun-Funn Adventure Park

NOTICE

- The adventure park charges admission for adults and children over 122 cm in height.
- Children less than 122 cm in height receive a 10% discount on their entry ticket.
- Toddlers less than 2 years old enter *Fun-Funn* for **free.**
- The park has restaurants and cafés. **No outside food or drink are allowed.**

Adults *each*	USD20
Children 2–12 *each*	USD10
Children under 2 *each*	Free

Have Fun-Funn fun!

Question 3

If there is space to show your working, it means you need to do some working out.

Write your answers in the correct spaces in the table.

a Work out the cost of entry for each member of Kwacy's family. Kwacy's little sister, Kalanda, is 1 m tall.

Family Members	Cost for Entry (USD)	Do your working out here
Steve	______	
Rebecca	______	
Kwacy	______	
Kalanda	______	

Check each person's age and height carefully. Kalanda is 1m tall – does she get the discount?

b Explain why Kwacy's entry fee is different from Kalanda's.

______

______

______

Question 4

Complete the invoice that the family received.

You may use the space below to show your working out.

Remember to:

Calculate subtotal first (add all items).

Then calculate tax (15% of subtotal).

Finally add subtotal + tax for total.

INVOICE

BILL FROM:
Fun-Funn Adventure Park Ltd.
One Love Boulevard, Mo Town Village
Mento Bay, Jamaica

BILL TO:
Steve Harris
76 Starfish Crescent
Super Town, Jamaica

INVOICE #: 2113021
INVOICE DATE: 20 December 2023
AMOUNT DUE: USD

Item	Description	Quantity	Unit Cost (USD)	Line Total (USD)
Adult	Entry		$20.00	
Child	Entry	1		$10.00
Child (<122 cm)	Entry			

Thank you for your business.
Have Fun-Funn fun!

SUBTOTAL	
TAX (15%)	
TOTAL	

Question 5

Kwacy's mom will pay the entry fee in Jamaican dollars (JMD). She has JMD 10 000 in cash in her purse. The exchange rate on the day they visit the park is USD 1 : JMD 157. Does she have enough cash or will she need to use her bank card? Show how you decided.

Exchange rate means US$1 = J$157. To convert US dollars to Jamaican: multiply by 157.

Part 3

Make sure you read the key and use the scale to make sense of the diagram.

The maze

After the adventure park, Kwacy and his family go to his cousin's birthday party. His cousin has made a maze of paths in the back yard. She has also placed obstacles along some of the paths to show where you cannot go through. The person who finds the shortest route through the maze will win a special prize. The diagram shows you a scaled plan of the maze.

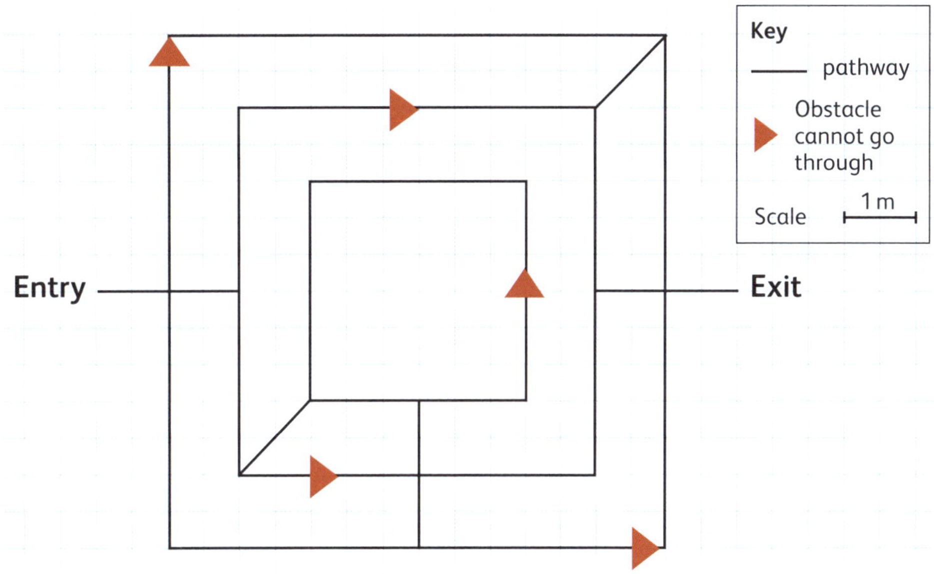

Question 6

Use the scale (1m) to measure distances along your path. Try different routes to find the shortest one.

a Estimate, in metres, the shortest possible distance from the entrance of the maze to the exit. Use your pencil to draw "x"s along the path you choose.

A path of 14m means you need 14 squares from entry to exit. Place obstacles to avoid blocking this path.

b Redesign the maze by rearranging all five of the obstacles in such a way that there is a clear path of at least 14 m from entrance to exit.

Use the same symbols as in the maze in the beginning of Part 3 or add a key to show what your symbols mean.

Add your obstacles to the pathway diagram below. Mark the clear path you have made on the diagram.

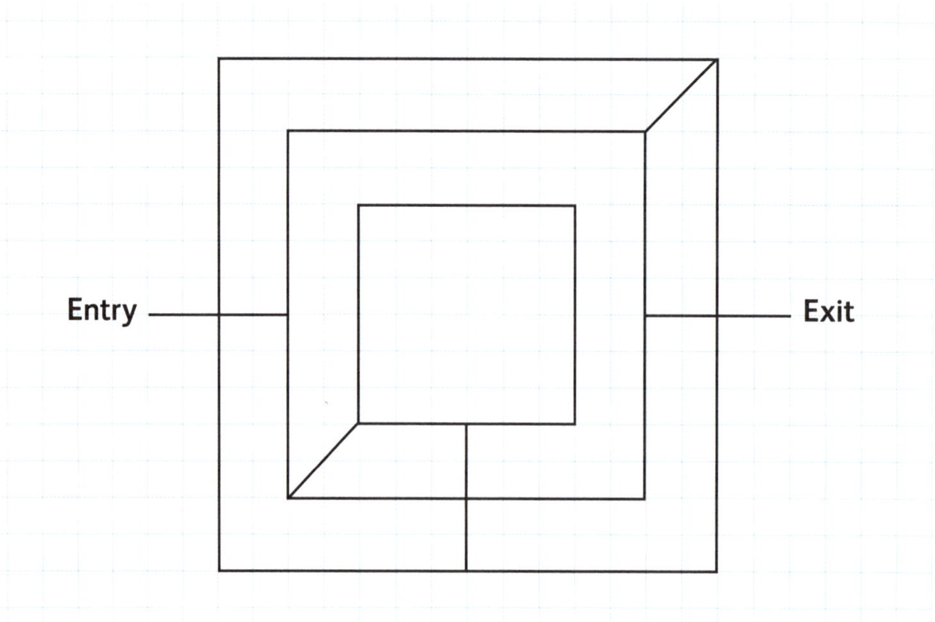

Part 4

A design is the same as a pattern. You can use numbers to describe many shape patterns.

Going home

Kwacy likes to solve puzzles in the car. His mom shows him a few designs. The first design is only 1 square. All other designs are made up of 1 square and some circles like this:

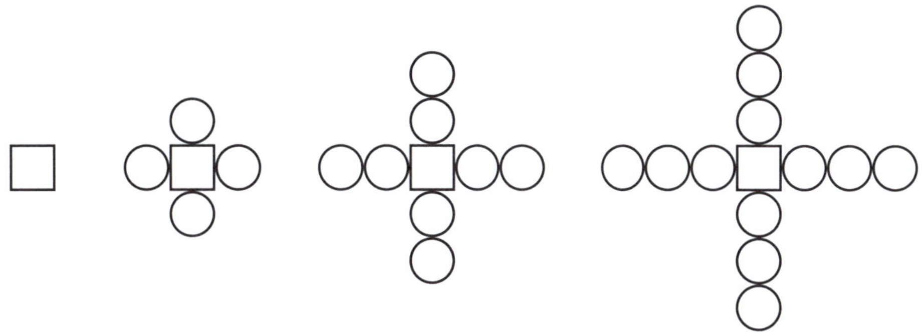

Design 1 **Design 2** **Design 3** **Design 4**

Question 7

Look at the pattern. Every design has 1 square. Count the circles in Designs 1–4 to find the pattern.

a How many squares and how many circles are there in Design 5?

b Draw Design 10 in the space below.

c Kwacy wonders if there is a way to work out the number of circles you need to make any design.

He makes this table to look for a pattern.

i Fill in the number of circles for Design 10.

Design	1	2	3	4	5	10	n
Number of circles	0	4	8	12	16	____	____

Look at the differences between consecutive numbers. 0, 4, 8, 12, 16. What number keeps getting added?

ii What number pattern can you see in the number of circles?

____

iii Kwacy notices that the number of circles is always four times one less than the design number. How can you write this mathematically?

Use your answer from iii to help you.

iv If n stands for any design number, what is the rule for finding the number of circles in Design n?

____

Use data that you know is correct to test your rule.

d Explain how you can check that your rule is correct.

____

____

e The teacher recorded how many circles were used in designs created by 10 students:

Number of circles: 8, 12, 16, 12, 20, 16, 24, 8, 16, 28

i What is the mean number of circles used?

____

ii What is the mode?

iii What is the range of circles used?

f (Optional challenge question):
Kwacy is building a pattern using stars.

- Design 1 has 5 stars.
- Each new design adds 3 more stars than the one before.

i How many stars are in Design 6?

ii Kwacy says one of his designs has exactly 50 stars. Is he correct? Explain your reasoning.

Reflection

Looking closely at your mistakes can help you learn and improve your work.

Think about any questions or parts of questions that you got wrong.

- Did you understand the question and know what to do? If not, what can you do about this?
- Did you make any mistakes in your calculations? If so, what can you do to avoid these mistakes in future?

Practice paper 3

This task has four parts:

- Part 1 has questions 1 and 2.
- Part 2 has question 3.
- Part 3 has questions 4 and 5.

Always check that you have everything you need before you start.

You will need a calculator, a pencil and an eraser.

Answer all the questions in each part in the spaces provided.

Remember to always show your working out.

→ Before you start ...

Some questions in the PEP tasks are open-ended. That means that there are different ways to answer and that there is more than one correct answer.

Find the questions that have a big working space.

Read each question. Think about how you can answer it to show your thinking and the strategies you used. Can you:

- do a labelled sketch?
- write a few sentences?
- draw a graph or make a table?
- write an equation and solve it?

At the pottery studio

Part 1

Taking a class

Question 1

David Pinto is a world-renowned potter who runs a studio in the Good Hope Estate near Falmouth. He offers classes for people who want to learn to make clay items.

Read this information about the classes offered and the prices.

Remember to highlight all the mathematical information so you can find it easily in order to answer questions.

I offer weekly classes in the school term, currently on Tuesday mornings from 10 a.m. to 12 a.m. It takes time and commitment to learn pottery and so these classes are ongoing for total beginners and those with any level of experience. Price depends on size of package signed up. Monthly $7 000 plus clay and firing.

For a one-time experience, you can request a private class. These cost US$270 for materials, instruction and firing (up to six people and $45 for each additional).

For experienced potters/artists, we host an annual 9-day international workshop. Next year it will take place from 17th to 26th May. Classes will be presented by visiting international artists and myself.

Costs (US dollars)
Registration fee: $500 (non-refundable and non-transferable)
Double room rate: $3 950 per person
Single room rate: $4 950 per person
Includes accommodation, meals and studio expense. Airfare is not included.
50% deposit is due by 15 October.
Balance is due by 15 January.

On average, we say there are four weeks in a month.

a Nadia signs up for weekly classes. She starts on 1 September and finishes at the end of January. What will this cost?

b A group of tourists decides to do a private class. There are eight persons in the group. Work out the cost of this class in US dollars.

c Shawn lives in Orlando in the USA. He has registered to attend the 9-day workshop and will be taking a single room. He books a return flight from Orlando to Kingston costing US$387.

i Calculate the total cost of the workshop for Shawn. Include the registration fee and the airfare.

Remember you can draw a rough number line to calculate with time. This shows your working.

ii The flight from Orlando to Kingston takes $6\frac{3}{4}$ hours. The time in Orlando is 1 hour ahead of the time in Kingston. If the flight takes off from Orlando at 10:20 a.m., what will the local time be when it lands in Kingston?

Speed = distance ÷ time

What is $2\frac{1}{2}$ hours as a decimal? (2.5 hours).

Now divide: 160 km ÷ 2.5 hours = ?

iii Shawn hires a car and drives from Normal Manley Airport in Kingston to the pottery studio. If it takes him $2\frac{1}{2}$ hours to make the 160-km journey, what is his average speed in kilometres per hour?

Question 2

When the students make table ware, they start with raw wet clay. The raw clay comes in 12.5-kg bags. David displays this poster to show students how much clay to use when they start making each type of item.

cup/mug 300 g

serving bowl 1 500 g

small bowl 450 g

side plate 1 000 g

soup bowl 900 g

serving plate 2 000 g

It is usually easier to convert measurements so that you are working in the same units.

a For each of these objects, calculate how many you can make from one full bag of clay, and how much clay would be left over.

i cup

ii soup bowls

iii serving bowls

First, work out: How much clay is needed for 1 cup + 1 small bowl? Then ask: How many of these pairs can you make from 12,500 g?

b From one bag of clay, a student decides to make equal numbers of cups and small bowls. How many of each can he make, and how much clay will he have left over?

c Suggest **two** different combinations of objects that a student could make that would use exactly **one** whole bag of clay, with none left over.

d Jay has 4 kg of raw clay. He wants to make six cups and six small bowls.

Nadia does a calculation and says that Jay does not have enough clay to make six cups and six bowls. Jay says it is possible if he makes the bowls slightly smaller.

Show how Jay could make six cups and six small bowls using only the 4 kg of clay that he has. The cups and bowls should all be the same size and he wants to use as much of the clay as possible.

You can show this in any way that makes sense, including using diagrams or tables.

Part 2

Learning more about clay

While the clay is still fresh and wet, the potter can shape and mould it. The wetter the clay, the softer it is. As it dries, it hardens. When most of the water has dried from the clay, the objects need to be fired in a kiln. This dries and hardens the material even more.

Question 3

a This table summarises the percentage of clay and water at different stages in the process. Some information is missing. Read the information in the boxes and use it to help you complete the table.

State of clay	Clay percentage (%)	Water percentage (%)
Slip	____________	____________
Plastic clay	____________	____________
____________	____________	10%
____________	98%	____________
____________	____________	0%

How do you convert fractions to equivalent percentages?

Mostly, you will work with clay in a **plastic state**. This is clay that is soft enough to mould. At this stage, it is about one-quarter water and three-quarters clay.

Slip is a mixture of half clay and half water. You use it like glue to stick one piece of clay to another.

As the clay dries out, its state changes to **leather hard**. When clay is leather hard, you can still smooth it off and carve into the surface, but if you try to change its shape, it will crack. Leather hard clay is about 1/10th water.

Use maths to give your reasons.

Finally, the clay is **bone-dry**. This means that it has almost entirely dry, and ready to be fired in the kiln.

When it is removed from the kiln, the clay is 100% dry. This state is known as **bisque**.

b Decide whether each statement is true or false based on the information in the table. Write true or false next to each statement and give a reason for your decision.

i Half a litre of slip contains about 1 cup of water. ______

______

ii 2 kg of slip contains about 1 kg of clay. ______

______

iii 10 cubic centimetres of leather hard clay contains about 1 cubic centimetre of water. ______

______

c A clay mug shrinks during drying and firing, losing 10% of its original size. The original height of the mug was 160 mm. What is the new height?

d The weight of a wet clay plate is 480 g. It loses 8% of its weight during drying and firing. What is the weight of the finished plate?

Part 3

Firing clay in the kiln

Question 4

During class, Nadia makes these notes about the temperature of the kiln.

- When I put the ceramics in, the kiln was at room temperature, which was 20°C.
- I turned on the kiln.
- After 1 hour, the temperature reached 250°C.
- It reached 700°C an hour later, and I reduced the temperature.
- The temperature dropped by 80°C within 30 minutes, and then another 120°C over the next 30 minutes.
- After four hours, the temperature was 410°C, and it dropped a hundred degrees an hour after that.
- I left the door closed for another five hours, and the temperature slowly came back down to room temperature before I opened it again.

Use a ruler and work as accurately as possible to draw the graph.

Use the labelled set of axes to plot a line graph to show how the temperature of the kiln changed from the start to the end of the firing time.

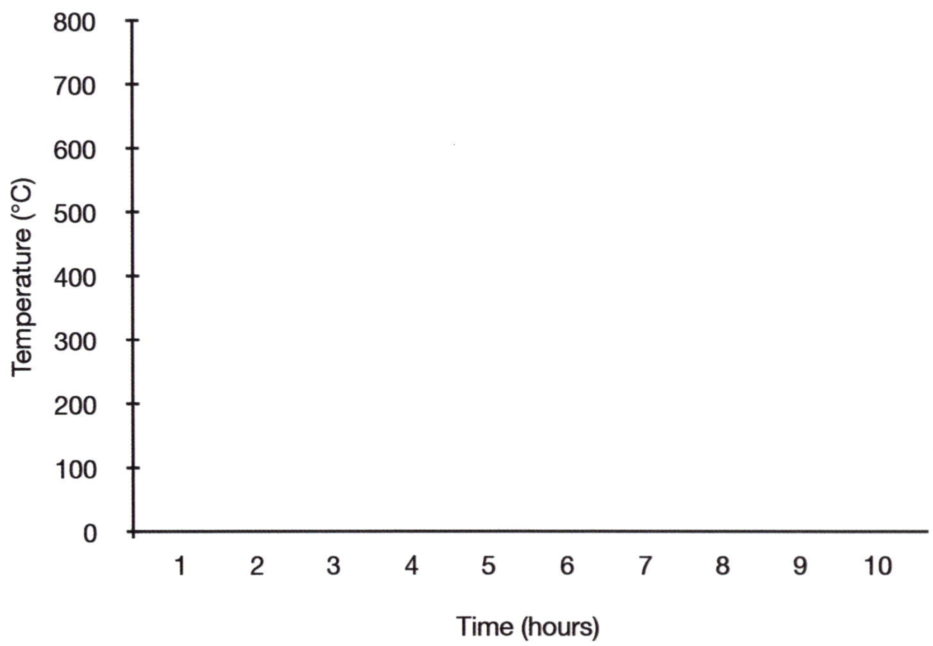

Question 5

Answer these questions about the graph.

a Give a reason **why** the temperature does not begin at zero, but the time does begin at zero.

____

____

Will a number line help you here?

b Nadia switched on the kiln at 8:45 a.m.

i At what time did it reach the maximum temperature?

____

ii What was the time when she opened the kiln after the firing?

____

c What is the range of temperatures shown on the graph?

____

When people use the word average, it usually refers to the mean.

d Nadia says that the temperature changed at an average rate of:

- 250 °C/hour in the first hour
- 350 °C in the first two hours
- 200 °C in the third hour.

Explain how all of these statements can be true.

____

____

____

____

____

e The kiln has a rectangular base measuring 80cm by 65cm, and it is 75cm tall.

i Calculate the area of the base of the kiln in square centimetres.

ii What is the volume of the kiln in cubic centimetres?

iii Convert your answer to part (ii) into cubic metres.

Reflection

Use your answers to the task to analyse your strengths and weaknesses.

- What areas are you good at?
- Which areas require more work?
- What work will you do?

Practice paper 4

This task has 3 parts:

- Part 1 has question 1.
- Part 2 has questions 2 and 3.
- Part 3 has question 4.

Always check that you have everything you need before you start.

You will need a calculator, a pencil and an eraser.

Answer all the questions in each part in the spaces provided.

Remember to always show your working out.

→ Before you start …

Skim read each question.

- Circle any number information in the question. Remember number information might be given in figures (5, $\frac{1}{2}$ or 1.2) or in words (four, three-quarters, nine kilometres).
- Underline key words in the question. These are clues that tell you what you need to do.
- Highlight any important information that is given in the question. You might need to use this to find the answer.

Part 1

Finding your way

Planning a route

Question 1

Always read the key and make sure you understand what is shown on the map before you start.

Always read the key and make sure you understand what is shown on the map before you start.

a The map shows the area around Seymour's home.

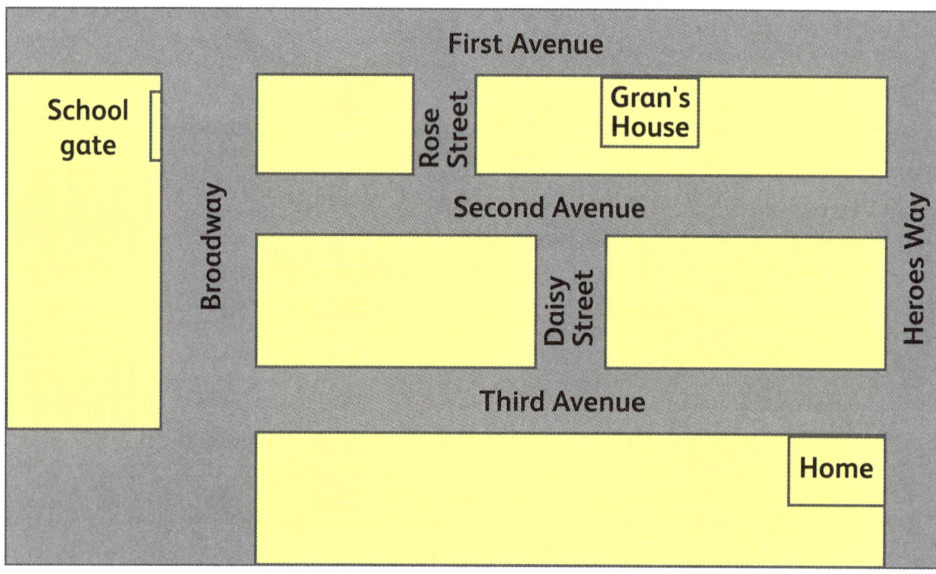

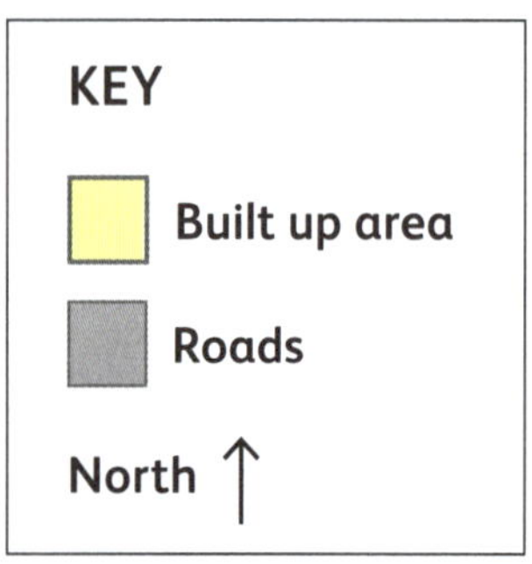

Seymour leaves school through the gate and walks home.

i Describe the shortest, most direct route that he can take.

ii On Tuesdays, Seymour goes to his gran's house on the way home. Draw a route he could take on the map and add it to the key.

iii On Fridays, Seymour likes to take a long walk home. What is the longest route he can take from school to home without backtracking? Draw your route on the map using a different colour or line.

What does "most direct" mean?

Work lightly in pencil if you need to try different routes. Make sure you clearly mark the final one you choose.

This map is not drawn to scale. You need to use the distances printed on it to work out the answers.

b Marcia is training to run a road race. She uses this map to plan her running routes around the neighbourhood.

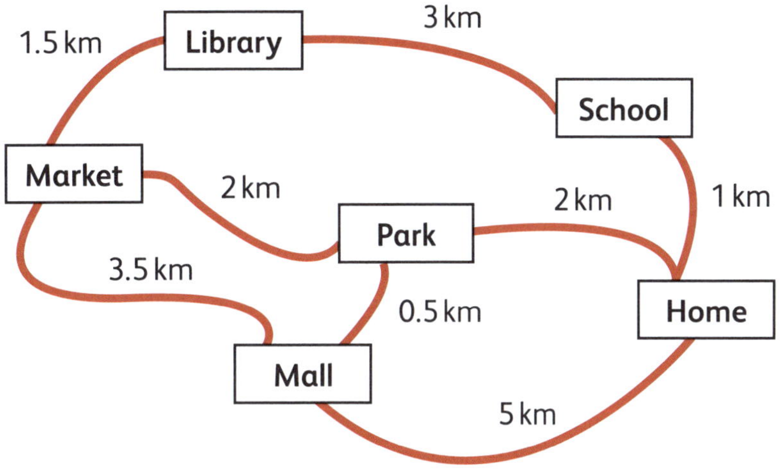

Which route (A, B, or C) is the longest?

Could you repeat part of a route to add more distance?

Could you combine two routes?

Test your idea: Add up all the distances to check if it's > 20 km.

Marcia's three favourite routes are:

Route A: Home → Park → Mall → Home

Route B: School → Library → Market → Park → Home

Route C: Mall → Market → Park → Home

i Which is the longest route?

____

ii If she runs each route twice during the week, how many kilometres will she run?

iii Marcia likes to run a route that is at least 20 km long on Saturdays. She starts and ends her run at home. What route could she take to do this?

____

Make sense of the map before you try to answer the questions.

c Portia uses a mapping app on her phone to get directions to walk from Lorna's Catering company to Devon House Bakery.

This is what her phone shows:

There are 5 options (skip A–E).

Find total distance for each.

Which equals 88 km?

i How far is the shortest route?

____

ii Which route takes $\frac{3}{4}$ of an hour?

____

iii Portia leaves Lorna's Catering at 10:36 a.m. and takes the route via Trafalgar Road. At what time should she reach Devon House Bakery?

Speed is given in kilometres per hour, so you need to think about minutes as fractions of an hour.

d There is a bus every 23 minutes on the Hope Road route. If a bus passes at 8:55 a.m., at what time will the next bus pass?

e It takes the bus six minutes to travel 3 km along Trafalgar Road. What is the average speed for this journey?

Part 2

Delivery drivers

Marcia's running routes use local roads. The same roads are used by delivery drivers. Let's look at how delivery routes are planned.

Matt is a delivery driver for a courier company. Every day, he collects packages from the main depot and delivers them to smaller depots where people can collect their parcels.

The diagram shows the location of the main depot and the five smaller depots. The distances between them are indicated on the diagram.

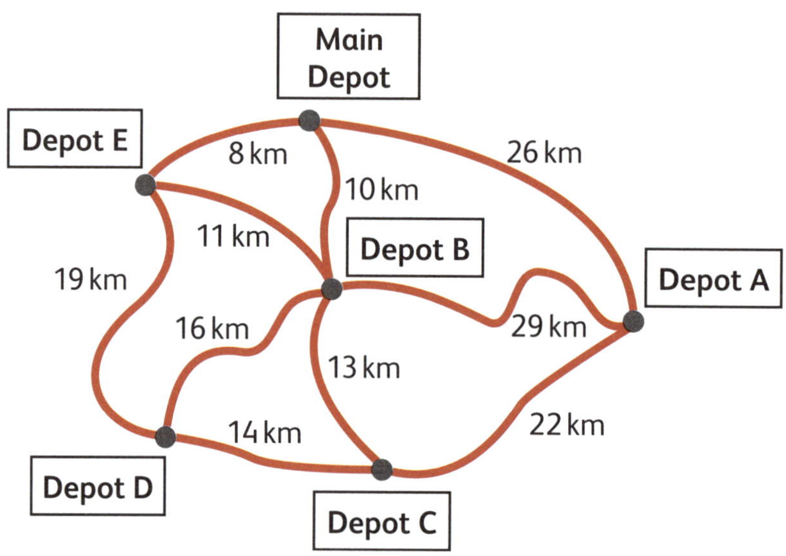

Question 2

a Matt plans to deliver packages to all five of the smaller depots. He only wants to drive to each depot once. One possible route is shown in the table.

Complete the table for the **shortest** route he can take.

Route	Kilometres travelled on each leg of the route	Total distance
MD → A → B → C → D → E → MD	26 + 29 + 13 + 14 + 19 + 8	109 km

Describe the route in the shortest form possible. Use the format in the table as a guideline.

b On Wednesday, he does not have to deliver any packages to Depot D. Describe the shortest route he can take and calculate how far he travels on that route.

What do you need to do to work this out? Show all the steps in your working.

c On Friday, he delivers to four out of the five depots covering 88 km. Work out his route and list the order in which he visited each depot.

d Matt recorded his daily mileage and package weights for one week:

Day	Distance (km)	Total package weight (kg)
Mon	152	236
Tue	173	308
Wed	141	204
Thus	164	268
Fri	201	340

i What is the mean weight of packages delivered per day?

ii What is the range of daily distances travelled?

iii On which day did Matt deliver the heaviest load?

Question 3

To plan efficient routes, the courier company tracks how many parcels they deliver each day for a month. They record the data on a stem-and-leaf plot.

Stem	Leaf
4	1 6
5	0 5 9
6	0 2 2 2 9
7	0 1 1 3 4 6
8	3 5 8
9	8 8 9
10	4 8
11	3 7
12	0 5

Key: 4 | 1 = 41

What value do the stems have? What value do the leaves have?

a The stem-and-leaf plot shows data for 28 days of the month. The number of parcels for the remaining three days of the month were seventy-eight, ninety-nine and one hundred twenty-seven.

Add this data to the stem-and-leaf plot.

b What is the range of the data?

____

c What is the mode of this data?

____

d The company says that on most days, they deliver more than 70 parcels. How can they tell this from the stem-and-leaf plot?

____

You are not being asked to do this calculation, you are being asked "how" to do it.

e Hiram wants to calculate the average number of parcels delivered each day. Write a clear set of instructions to show how he could use the data in the stem-and-leaf plot to work that out.

____

____

____

____

Part 3

Calculating what you will pay

Question 4

Patrice needs a new laptop to use navigation and mapping apps for his delivery work. He's found one locally for J$39,500 but wonders if buying online from an international site might be cheaper once shipping and customs are included.

Patrice finds the same laptop model on an international shopping website for US$198.

a Using an exchange rate of US$1 to J$154, calculate the cost of the laptop on the online shopping site.

Use the value you already worked out for the laptop.

b When Patrice fills in his details and address, the shopping site automatically adds import duties and shipping of US$71.78 to the order.

If Patrice goes ahead with the purchase, what will he end up paying for the laptop in Jamaican dollars?

This is a multi-step problem. Read it carefully and set out each step clearly.

c The total of US$71.78 includes US$12 for shipping, GST on the price of the laptop and customs charges to import the laptop into Jamaica. Customs charges are a percentage of the price of the item excluding GST.

Work out the percentage charged by customs and give your answer to the nearest tenth of a per cent. Show all the steps in the working.

d How do you think it is possible for the company in Jamaica to sell the same laptop for less than Patrice would end up paying? Suggest **two** ways.

e (Optional challenge question):
Patrice finds the same laptop on a Canadian website for CA$285, with shipping and import fees of CA$55. The exchange rate is CA$1 = J$112.

i Calculate the total cost in Jamaican dollars if Patrice buys from Canada.

ii Which is cheaper: buying from the USA or Canada? By how much?

iii The exchange rate changes to CA$1 = J$120. Would this make the Canadian option cheaper or more expensive than the USA option? Show your working.

Reflection

You can get good marks in mathematics by clearly showing your working out and methods even if you get the final answer incorrect.

Evaluate your answers to the task.

- Did you show your working?
- Is your working clearly laid out and systematic?
- Does your working show that you understood the question and give evidence that you know how to answer it?
- Do you think you would get part marks for any questions where you got the incorrect answer? Support your opinion.

Practice paper 5

This task has 3 parts:

- Part 1 has question 1.
- Part 2 has questions 2 and 3.
- Part 3 has questions 4 and 5.

Always check that you have everything you need before you start.

You will need a calculator, a pencil and an eraser.

Answer all the questions in each part in the spaces provided.

Remember to always show your working out.

→ Before you start ...

When you do a task or test, you can answer the questions in any order.

It is better to do the questions that are easy first. That way you have more time to spend on the more challenging parts of the task.

This task has five questions.

Skim read the questions and decide on an order to answer them.

Write your order here:

First, I will do question ______________.

Then, I will do question ______________.

Next, I will finish question ______________.

After that, I will try question ______________.

Lastly, I will do question ______________.

Give **one** reason why you chose this order. Write it here.

____

A school trip

Part 1

Booking flights

A cricket team from Montego Bay are going to travel to Port of Spain in Trinidad and Tobago to take part in a competition. The team will fly from Kingston to Port of Spain. Ms George is in charge of booking flights for the trip. She makes these notes about the flights.

Why do you think you are given this information?

Direct flights
Distance 1 860 kilometres
Flying time 2 hours 45 minutes
US$428 per person return
Exchange rate US$1 : J$154.75

Kingston to Port of Spain
Depart 4:50 p.m., Arrive 8:35 p.m.
Port of Spain to Kingston
Depart 8 a.m.

Question 1

Where will you find the information you need to answer these questions?

a 15 players and two coaches will go on the trip. Calculate the total cost of the tickets in Jamaican dollars.

b The flight time is 2 hours 45 minutes. How many minutes is this in total?

i If the flight takes 2 hours and 45 minutes, what speed does the plane fly at? Give your answer to the nearest whole kilometre.

Draw a timeline from 4:50 p.m. to 8:35 p.m. Count the hours and minutes carefully.

c The flight leaves Kingston at 4:50 p.m. and arrives in Port of Spain at 8:35 p.m. Cassandra says that the flight must be 3 hours and 45 minutes long. Explain why Cassandra is incorrect.

____

____

d The flight from Port of Spain back to Kingston leaves at 8 a.m.

i The team must be at the airport two hours before their flight and the minibus trip from their accommodation to the airport takes 20 minutes. At what time do they need to leave their accommodation to be at the airport two hours before they fly?

____

____

ii The flight leaves ten minutes late but it still takes 2 hours and 45 minutes to reach Kingston. At what time will the plane land? Give your answer in local time.

____

____

Part 2

Remember, you can highlight information to make it easier to find.

Baggage allowances

Read the information about baggage allowances carefully before answering the questions.

Sporting equipment

Certain types of sporting equipment may form part of your free checked baggage allowance. The maximum weight allowed per item is 23 kg (50 lb) with maximum linear dimensions of (H + W + L) 157 cm (62 in). Oversized sporting equipment will be permitted as excess baggage on a standby basis, at the excess baggage charge of US$6.50/kg.

We accept bats, wickets, pads and balls as cricket equipment as long as these items are stored in a recognised cricket coffin or soft team bag.

Carry-on baggage

Each passenger may carry **1** carry-on bag and **1** personal item. Handbags, small backpacks, briefcases and laptop bags are classified as personal items.

All items must comply with these guidelines.

Make sure you know what each diagram is showing you.

Carry-on bags

Max weight 10 kg

Max size

L56 cm × W35 cm × H23 cm

Note 2.54 cm = 1 in

Personal items

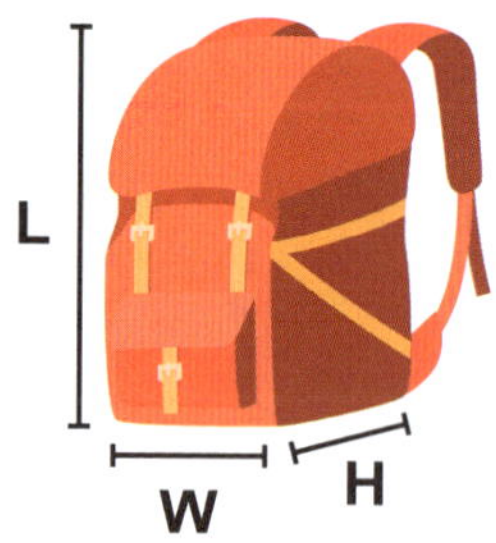

Max size

L43 cm × W33 cm × H15 cm

Checked bags

1 per customer free

Max weight 23 kg

Max linear dimensions (L + W + H) = 157 cm

Note 2.54 cm = 1 in

Question 2

Convert inches to cm: multiply by 2.54.

a The school has cricket coffin bags that are 38 in long, 14 in wide and 14.5 in high. Will these be suitable for the trip?

b The team also has duffel bags with dimensions 77.5 × 36 × 42 cm. Are these acceptable kit bags?

c Do the carry-on bag measurements include the handles and wheels? Explain your answer.

d What is the maximum volume of a personal item?

You are told that her reasoning is wrong. You have to explain what mistake she has made.

e Sandra has a carry-on bag that is 24 cm high, 30 cm wide and 60 cm tall. She says it will be allowed because 24 + 30 + 60 = 114 and the airline measurements are 56 + 35 + 23 = 114. What is wrong with Sandra's reasoning?

f The coach has a new suitcase. The height is three times the width and the length is 20 cm less than the height. Work out the greatest value that the width can be to allow this as checked baggage if it weighs under 23 kg when it is packed.

Question 3

Where can you find the information you need to answer these questions?

a When they get to the airport, the coach finds that his luggage is 5.8 kg over the limit. What will he have to pay for excess baggage?

b Look at Josh's baggage. Will he be able to carry this with him without paying any penalties? Show how you work this out and give reasons for your answers.

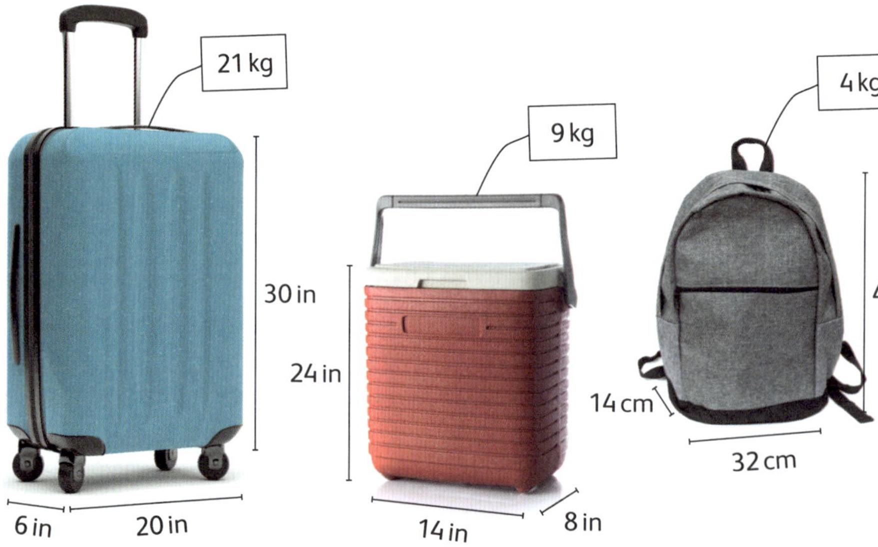

c The masses of the players and the coaches with their hand baggage are:

55	61	65	57	53
47	54	58	60	51
75	54	53	61	68
83	52			

The checked bags for the whole team come to 401 kg and the team has 3 kit bags weighing 22, 23 and 21 kg.

i Which two masses do you think are the coaches? Why?

ii What is the average mass of a player with hand baggage?

iii How much excess baggage charges did the team and coaches have to pay?

iv What is the total mass of the kit?

Part 3

On the other side

Question 4

Before the team leaves, Micah checks the weather forecast for their week in Port of Spain so they know what to pack.

Mon	Tue	Wed	Thu	Fri	Sat	Sun
31 °C	26 °C	26 °C	27 °C	28 °C	29 °C	31 °C

a What range of temperature is predicted?

____

b What is the mean temperature likely to be during their visit?

____

c Describe the chance of a game being rained out if the games take place from Tuesday to Saturday.

____

____

Remember, chance is the same as probability and you can describe it on a scale of 0 to 1.

Question 5

On the return flight, Vishal notices the plane descending. The pilot explains that planes follow this average descent pattern when landing in Kingston.

"Descending" means going down.

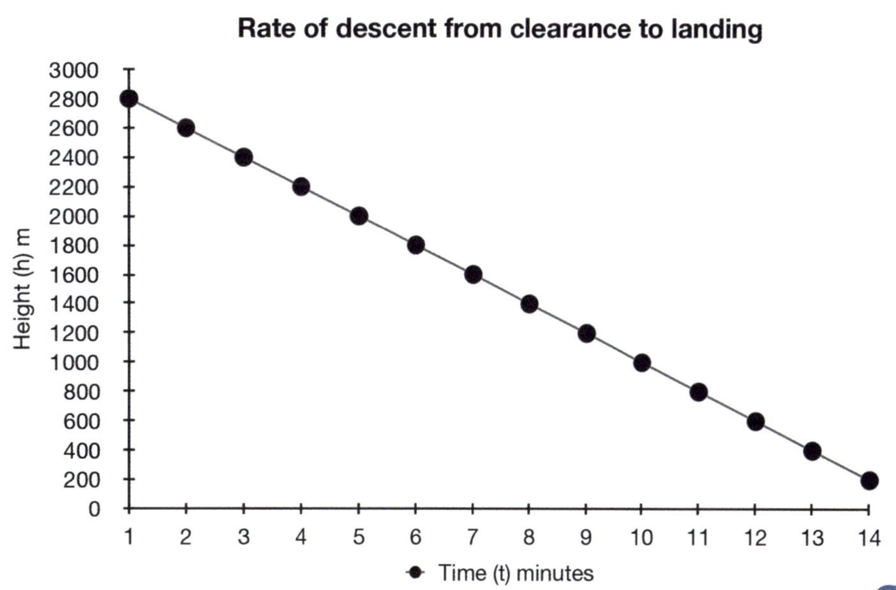

A rule is another word for formula.

a Describe the trend shown on the graph.

____

b Write a rule for calculating the height of the plane at any time during its descent.

____

c Use your rule to find:

i the height of the plane 4 minutes into its descent

____

ii the time it takes for the plane to descend 675 metres

____

iii the height 150 s before the plane touches down on the runway.

____

Reflection

Make a list of three things in this task that you did well.

____

____

____

- What could you do better?
- Write down one thing that you learned in another task that helped you here.

____

____

Practice paper 6

This task has 3 parts:

- Part 1 has question 1.
- Part 2 has questions 2 and 3.
- Part 3 has questions 4 and 5.

Always check that you have everything you need before you start.

You will need a calculator, a pair of compasses, a pencil and an eraser.

Answer all the questions in each part in the spaces provided.

Remember to always show your working out.

→ Before you start …

You have learned a lot about doing tasks in this book.

Write down **three** important things that you should always do before you start a task.

1 ______________________________

2 ______________________________

3 ______________________________

Why is it important to do these things?

____

____

Kingston National Gallery

Class 6 is learning about Jamaican culture and heritage. They decide to find out more about the National Gallery of Jamaica.

Part 1

Before you go

Read this information about Kingston National Gallery before you answer the questions.

KINGSTON NATIONAL GALLERY

Kingston National Gallery was established in 1974.
In July 2014, a satellite branch – National Gallery West – was opened in Montego Bay.

OPENING HOURS

Tuesday to Thursday: 10 a.m. – 4:30 p.m.

Friday: 10 a.m. – 4 p.m.

Saturday: 10 a.m. – 3 p.m.

Sunday and Monday: Closed

TICKET PRICES

Adults: $400

Senior Citizens (65+): $200

Under 16: Free (ID needed)

Question 1

Read the first paragraph carefully. Look for two place names.

a Where are the two branches of the National Gallery located?

____

You are being asked to give an answer in units of time here.

b How long has the National Gallery been in existence?

____

c How many years after the establishment of the National Gallery did the National Gallery West open?

____

Show the steps you follow to work this out. Do not just write an answer.

d For how many hours a week is the National Gallery in Kingston open?

Adults × $400 = ? Senior citizens × $200 = ?

Children under 16 = Free

Add the totals

e Sharyn and her family visit the National Gallery on Wednesday. They arrive at 11:45 a.m. They get entry tickets for three adults, two senior citizens and two children ages 14 and 9.

i What will the tickets cost for the whole family?

They arrive at 11:45 a.m. Gallery closes at 4:30 p.m. on Wednesday.

ii How long can the family spend in the gallery?

Part 2

At the National Gallery

Question 2

This is the logo of the National Gallery.

Think about the maths words you can use to describe the parts of the logo.

a Describe the N in the logo using mathematical terms.

____

____

b Describe the G in the logo using the correct names of circle parts.

____

____

Use your compass to draw the circle parts.

c Greyton has started to draw the logo as a set of lines on a grid. Use what you know about the properties of shapes to complete the logo as accurately as you can.

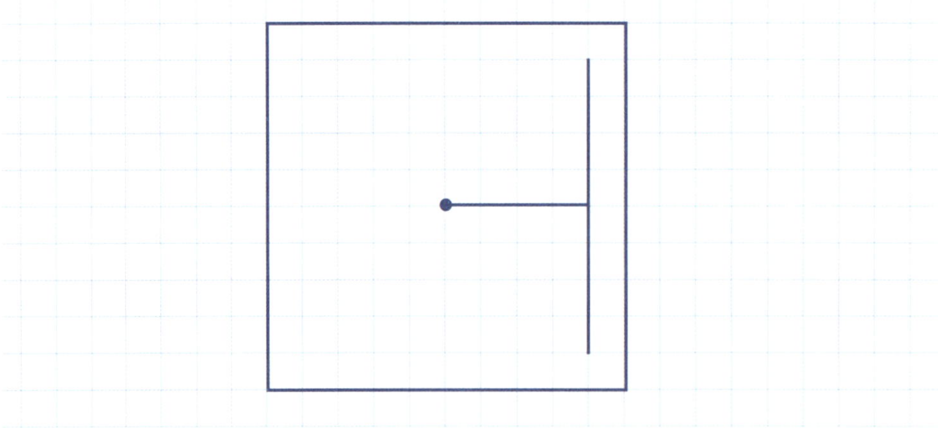

Question 3

Use information from the photograph to help you answer the questions.

a Sharyn walks from the lower level up two sets of eight stairs to the landing area, then she walks up the last set of eight stairs to the upper level. If each stair is 15 cm high, what is the difference in height between the lower and upper levels?

b The light panels on the ceiling are 2 m long and 1 m wide. What is the combined area of the light panels you can see in the photograph?

Count carefully, but also say how many you have counted so you can get credit for working even if you miss out some light panels.

You can draw on the photo if it will help you.

c The stairways are 1.8 m wide. The wall between the stairways on the landing area is 3.6 m long. When Sharyn reaches the upper level, she turns left. What is the length of the barrier fence along that section of the gallery?

Part 3

Maths and art

Question 4

The National Gallery offers art classes for students. At a painting class, Desray painted this work.

Desray says the circles are congruent with a radius of 7 cm in a frame with right-angled corners.

What does Desray mean? Can you mark this on the picture?

a What can you say about the shape of the red frame? Why?

____

____

Will it help if you draw a rough sketch of this on the diagram?

b Desray decides to add a triangle to the design. She marks the centre of each circle and joins the three points to make a triangle. What is the perimeter of the triangle she draws?

Flat shapes like this have two dimensions– length and width. You must work out both of these for each option.

Question 5

The gallery wants to print some invitations to an exhibition. They will print one large invitation (30 cm × 20 cm) and two smaller invitations on one sheet of paper. The printer shows them two different options for doing this.

Work out the dimensions of the smaller invitations for each option. Show how you do this.

Option A

Option B

Reflection

Can you answer "yes" to all these questions?

If not, write down what you need to do to change your answer.

- Did I understand all the questions?
- Did I know what to do to answer them?
- Did I follow the instructions correctly?
- Was my work clear and organised?
- Did I perform procedures accurately?
- Was I able to use my calculator efficiently?
- Did I explain and justify my answers when asked to do so?
- Were my solutions correct?